Grundlehren der mathematischen Wissenschaften 311

A Series of Comprehensive Studies in Mathematics

Editors

M. Artin S.S. Chern J. Coates J.M. Fröhlich
H. Hironaka F. Hirzebruch L. Hörmander C.C. Moore
J.K. Moser M. Nagata W. Schmidt D.S. Scott
Ya.G. Sinai J. Tits M. Waldschmidt S. Watanabe

Managing Editors

M. Berger B. Eckmann S.R.S. Varadhan

Springer
*Berlin
Heidelberg
New York
Barcelona
Budapest
Hong Kong
London
Milan
Paris
Santa Clara
Singapore
Tokyo*

Mariano Giaquinta
Stefan Hildebrandt

Calculus of Variations II

The Hamiltonian Formalism

With 82 Figures

 Springer

Mariano Giaquinta
Università di Firenze, Dipartimento di Matematica Applicata "G. Sansone"
Via S. Marta 3, I-50139 Firenze, Italy

Stefan Hildebrandt
Universität Bonn, Mathematisches Institut
Wegelerstr. 10, D-53115 Bonn, Germany

Mathematics Subject Classification: 49-XX, 53-XX, 70-XX

ISBN 3-540-57961-3 Springer-Verlag Berlin Heidelberg New York

Library of Congress Cataloging-in-Publication Data. Giaquinta, Mariano, 1947– . Calculus of variations/Mariano Giaquinta, Stefan Hildebrandt. p. cm.–(Grundlehren der mathematischen Wissenschaften; 310–311) Includes bibliographical references and indexes. Contents: 1. The Lagrangian formalism – 2. The Hamiltonian formalism. ISBN 3-540-50625-X (Berlin: v. 1). – ISBN 0-387-50625-X (New York: v. 1). – ISBN 3-540-57961-3 (Berlin: v. 2). – ISBN 0-387-57961-3 (New York: v. 2) 1. Calculus of variations. I. Hildebrandt, Stefan. II. Title. III. Series. QA315.G46 1996 515'.64–dc20 96-20429

This work is subject to copyright. All rights are reserved, whether the whole or part of the material is concerned, specifically those of translation, reprinting, reuse of illustrations, recitation, broadcasting, reproduction on microfilms or in other way, and storage in data banks. Duplication of this publication or parts thereof is permitted only under the provisions of the German Copyright Law of September 9, 1965, in its current version, and a copyright fee must always be obtained from Springer-Verlag. Violations fall under the prosecution act of the German Copyright Law.

© Springer-Verlag Berlin Heidelberg 1996
Printed in Germany

Cover design: Springer-Verlag, Design & Production

Typesetting: Asco Trade Typesetting Ltd., Hong Kong

SPIN: 10128795   41/3140/SPS – 5 4 3 2 1 0 – Printed on acid-free paper

Preface

This book describes the classical aspects of the variational calculus which are of interest to analysts, geometers and physicists alike. Volume 1 deals with the formal apparatus of the variational calculus and with nonparametric field theory, whereas Volume 2 treats parametric variational problems as well as Hamilton–Jacobi theory and the classical theory of partial differential equations of first order. In a subsequent treatise we shall describe developments arising from Hilbert's 19th and 20th problems, especially direct methods and regularity theory.

Of the classical variational calculus we have particularly emphasized the often neglected theory of inner variations, i.e. of variations of the independent variables, which is a source of useful information such as monotonicity formulas, conformality relations and conservation laws. The combined variation of dependent and independent variables leads to the general conservation laws of Emmy Noether, an important tool in exploiting symmetries. Other parts of this volume deal with Legendre–Jacobi theory and with field theories. In particular we give a detailed presentation of one-dimensional field theory for nonparametric and parametric integrals and its relations to Hamilton–Jacobi theory, geometrical optics and point mechanics. Moreover we discuss various ways of exploiting the notion of convexity in the calculus of variations, and field theory is certainly the most subtle method to make use of convexity. We also stress the usefulness of the concept of a null Lagrangian which plays an important role in several instances. In the final part we give an exposition of Hamilton–Jacobi theory and its connections with Lie's theory of contact transformations and Cauchy's integration theory of partial differential equations.

For better readability we have mostly worked with local coordinates, but the global point of view will always be conspicuous. Nevertheless we have at least once outlined the coordinate-free approach to manifolds, together with an outlook onto symplectic geometry.

Throughout this volume we have used the classical *indirect method* of the calculus of variations solving first Euler's equations and investigating thereafter which solutions are in fact minimizers (or maximizers). Only in Chapter 8 we have applied direct methods to solve minimum problems for parametric integrals. One of these methods is based on results of field theory, the other uses the concept of lower semicontinuity of functionals. *Direct methods* of the calculus of variations and, in particular, existence and regularity results for minimizers of multiple integrals will be subsequently presented in a separate treatise.

We have tried to write the present book in such a way that it can easily be read and used by any graduate student of mathematics and physics, and by nonexperts in the field. Therefore we have often repeated ideas and computations if they appear in a new context. This approach makes the reading occasionally somewhat repetitious, but the reader has the advantage to see how ideas evolve and grow. Moreover he will be able to study most parts of this book without reading all the others. This way a lecturer can comfortably use certain parts as text for a one-term course on the calculus of variations or as material for a reading seminar.

We have included a multitude of examples, some of them quite intricate, since examples are the true lifeblood of the calculus of variations. To study specific examples is often more useful and illustrative than to follow all ramifications of the general theory. Moreover the reader will often realize that even simple and time-honoured problems have certain peculiarities which make it impossible to directly apply general results.

In the *Scholia* we present supplementary results and discuss references to the literature. In addition we present historical comments. We have consulted the original sources whenever possible, but since we are no historians we might have more than once erred in our statements. Some background material as well as hints to developments not discussed in our book can also be found in the *Supplements*.

A last word concerns the size of our project. The reader may think that by writing two volumes about the classical aspects of the calculus of variations the authors should be able to give an adequate and complete presentation of this field. This is unfortunately not the case, partially because of the limited knowledge of the authors, and partially on account of the vast extent of the field. Thus the reader should not expect an encyclopedic presentation of the entire subject, but merely an introduction in one of the oldest, but nevertheless very lively areas of mathematics. We hope that our book will be of interest also to experts as we have included material not everywhere available. Also we have examined an extensive part of the classical theory and presented it from a modern point of view.

It is a great pleasure for us to thank friends, colleagues, and students who have read several parts of our manuscript, pointed out errors, gave us advice, and helped us by their criticism. In particular we are very grateful to Dieter Ameln, Gabriele Anzellotti, Ulrich Dierkes, Robert Finn, Karsten Große-Brauckmann, Anatoly Fomenko, Hermann Karcher, Helmut Kaul, Jerry Kazdan, Rolf Klötzler, Ernst Kuwert, Olga A. Ladyzhenskaya, Giuseppe Modica, Frank Morgan, Heiko von der Mosel, Nina N. Uraltseva, and Rüdiger Thiele. The latter also kindly supported us in reading the galley proofs. We are much indebted to Kathrin Rhode who helped us to prepare several of the examples. Especially we thank Gudrun Turowski who read most of our manuscript and corrected numerous mistakes. Klaus Steffen provided us with example $\boxed{0}$ in 3,*1* and the regularity argument used in 3,*6* nr. 11. Without the patient and excellent typing and retyping of our manuscripts by Iris Pützer and Anke Thiedemann this book could not have been completed, and we appreciate their invaluable help as well as the patience of our Publisher and the constant and friendly encouragement by Dr. Joachim Heinze. Last but not least we would like to extend our thanks to *Consiglio Nazionale delle Ricerche*, to *Deutsche Forschungsgemeinschaft*, to *Sonderforschungsbereich* 256 *of Bonn University*, and to the *Alexander von Humboldt Foundation*, which have generously supported our collaboration.

Bonn and Firenze, February 14, 1994　　　　　　　　　　Mariano Giaquinta
　　　　　　　　　　　　　　　　　　　　　　　　　　Stefan Hildebrandt

Contents of Calculus of Variations I and II

Calculus of Variations I: The Lagrangian Formalism

Introduction
Table of Contents

Part I.   The First Variation and Necessary Conditions
      Chapter 1.   The First Variation
      Chapter 2.   Variational Problems with Subsidiary Conditions
      Chapter 3.   General Variational Formulas

Part II. The Second Variation and Sufficient Conditions
      Chapter 4.   Second Variation, Excess Function, Convexity
      Chapter 5.   Weak Minimizers and Jacobi Theory
      Chapter 6.   Weierstrass Field Theory for One-dimensional Integrals and Strong Minimizers

Supplement. Some Facts from Differential Geometry and Analysis
A List of Examples
Bibliography
Index

Calculus of Variations II: The Hamiltonian Formalism

Table of Contents

Part III. Canonical Formalism and Parametric Variational Problems
      Chapter 7.   Legendre Transformation, Hamiltonian Systems, Convexity, Field Theories
      Chapter 8.   Parametric Variational Integrals

Part IV. Hamilton-Jacobi Theory and Canonical Transformations
      Chapter 9.   Hamilton-Jacobi Theory and Canonical Transformations
      Chapter 10. Partial Differential Equations of First Order and Contact Transformations

A List of Examples
A Glimpse at the Literature
Bibliography
Index

Introduction

The Calculus of Variations is the art to find optimal solutions and to describe their essential properties. In daily life one has regularly to decide such questions as which solution of a problem is best or worst; which object has some property to a highest or lowest degree; what is the optimal strategy to reach some goal. For example one might ask what is the shortest way from one point to another, or the quickest connection of two points in a certain situation. The isoperimetric problem, already considered in antiquity, is another question of this kind. Here one has the task to find among all closed curves of a given length the one enclosing maximal area. The appeal of such optimum problems consists in the fact that, usually, they are easy to formulate and to understand, but much less easy to solve. For this reason the calculus of variations or, as it was called in earlier days, the isoperimetric method has been a thriving force in the development of analysis and geometry.

An ideal shared by most craftsmen, artists, engineers, and scientists is the principle of the economy of means: What you can do, you can do simply. This aesthetic concept also suggests the idea that nature proceeds in the simplest, the most efficient way. Newton wrote in his *Principia*: "*Nature does nothing in vain, and more is in vain when less will serve; for Nature is pleased with simplicity and affects not the pomp of superfluous causes.*" Thus it is not surprising that from the very beginning of modern science optimum principles were used to formulate the "laws of nature", be it that such principles particularly appeal to scientists striving toward unification and simplification of knowledge, or that they seem to reflect the preestablished harmony of our universe. Euler wrote in his *Methodus inveniendi* [2] from 1744, the first treatise on the calculus of variations: "*Because the shape of the whole universe is most perfect and, in fact, designed by the wisest creator, nothing in all of the world will occur in which no maximum or minimum rule is somehow shining forth.*" Our belief in the best of all possible worlds and its preestablished harmony claimed by Leibniz might now be shaken; yet there remains the fact that many if not all laws of nature can be given the form of an extremal principle.

The first known principle of this type is due to Heron from Alexandria (about 100 A.D.) who explained the law of reflection of light rays by the postulate that *light must always take the shortest path*. In 1662 Fermat succeeded in deriving the law of refraction of light from the hypothesis that *light always propagates in the quickest way from one point to another*. This assumption is now called *Fermat's principle*. It is one of the pillars on which geometric optics rests; the other one is *Huygens's principle* which was formulated about 15 years later. Further, in his letter to De la Chambre from January 1, 1662, Fermat motivated his principle by the following remark: "*La nature agit toujour par les voies les plus courtes.*" (Nature always acts in the shortest way.)

About 80 years later Maupertuis, by then President of the Prussian Academy of Sciences, resumed Fermat's idea and postulated his metaphysical principle of the *parsimonious universe*, which later became known as "*principle of least action*" or "*Maupertuis's principle*". He stated: *If there occurs some change in nature, the amount of action necessary for this change must be as small as possible.*

"Action" that nature is supposed to consume so thriftily is a quantity introduced by Leibniz which has the dimension "energy × time". It is exactly that quantity which, according to Planck's quantum principle (1900), comes in integer multiples of the elementary quantum h.

In the writings of Maupertuis the action principle remained somewhat vague and not very convincing, and by Voltaire's attacks it was mercilessly ridiculed. This might be one of the reasons why Lagrange founded his *Méchanique analitique* from 1788 on d'Alembert's principle and not on the least action principle, although he possessed a fairly general mathematical formulation of it already in 1760. Much later Hamilton and Jacobi formulated quite satisfactory versions of the action principle for point mechanics, and eventually Helmholtz raised it to the rank of the most general law of physics. In the first half of this century physicists seemed to prefer the formulation of natural laws in terms of space–time differential equations, but recently the principle of least action had a remarkable comeback as it easily lends itself to a global, coordinate-free setup of physical "field theories" and to symmetry considerations.

The development of the calculus of variations began briefly after the invention of the infinitesimal calculus. The first problem gaining international fame, known as "problem of quickest descent" or as "brachystochrone problem", was posed by Johann Bernoulli in 1696. He and his older brother Jakob Bernoulli are the true founders of the new field, although also Leibniz, Newton, Huygens and l'Hospital added important contributions. In the hands of Euler and Lagrange the calculus of variations became a flexible and efficient theory applicable to a multitude of physical and geometric problems. Lagrange invented the δ-calculus which he viewed to be a kind of "higher" infinitesimal calculus, and Euler showed that the δ-calculus can be reduced to the ordinary infinitesimal calculus. Euler also invented the multiplier method, and he was the first to treat variational problems with differential equations as subsidiary conditions. The development of the calculus of variations in the 18th century is described in the booklet by Woodhouse [1] from 1810 and in the first three chapters of H.H. Goldstine's historical treatise [1]. In this first period the variational calculus was essentially concerned with deriving necessary conditions such as Euler's equations which are to be satisfied by minimizers or maximizers of variational problems. Euler mostly treated variational problems for single integrals where the corresponding Euler equations are ordinary differential equations, which he solved in many cases by very skillful and intricate integration techniques. The spirit of this development is reflected in the first parts of this volume. To be fair with Euler's achievements we have to emphasize that he treated in [2] many more one-dimensional variational problems than the reader can find anywhere else including our book, some of which are quite involved even for a mathematician of today.

However, no *sufficient conditions* ensuring the minimum property of solutions of Euler's equations were given in this period, with the single exception of a paper by Johann Bernoulli from 1718 which remained unnoticed for about 200 years. This is to say, analysts were only concerned with determining solutions of Euler equations, that is, with stationary curves of one-dimensional variational problems, while it was more or less taken for granted that such stationary objects furnish a real extremum.

The sufficiency question was for the first time systematically tackled in Legendre's paper [1] from 1788. Here Legendre used the idea to study the second variation of a functional for deciding such questions. Legendre's paper contained some errors, pointed out by Lagrange in 1797, but his ideas proved to be fruitful when Jacobi resumed the question in 1837. In his short paper [1] he sketched an entire theory of the second variation including his celebrated theory of conjugate points, but all of his results were stated with essentially no proofs. It took a whole generation of mathematicians to fill in the details. We have described the basic features of the *Legendre–Jacobi theory* of the second variation in Chapters 4 and 5 of this volume.

Euler treated only a few variational problems involving multiple integrals. Lagrange derived the "Euler equations" for double integrals, i.e. the necessary differential equations to be satisfied by minimizers or maximizers. For example he stated the minimal surface equation which characterizes the stationary surface of the nonparametric area integral. However he did not indicate how one can obtain solutions of the minimal surface equation or of any other related Euler equation. Moreover neither he nor anyone else of his time was able to derive the *natural boundary conditions* to be satisfied by, say, minimizers of a double integral subject to free boundary conditions since the tool of "integration by parts" was not available. The first to successfully tackle two-dimensional variational problems with free boundaries was Gauss in his paper [3] from 1830 where he established a variational theory of capillary phenomena based on Johann Bernoulli's *principle of virtual work* from 1717. This principle states that in equilibrium no work is needed to achieve an infinitesimal displacement of a mechanical system. Using the concept of a potential energy which is thought to be attached to any state of a physical system, Bernoulli's principle can be replaced by the following hypothesis, *the principle of minimal energy*: The equilibrium states of a physical system are stationary states of its potential energy, and the stable equilibrium states minimize energy among all other "virtual" states which lie close-by.

For capillary surfaces not subject to any gravitational forces the potential energy is proportional to their surface area. This explains why the phenomenological theory of soap films is just the theory of surfaces of minimal area.

After Gauss free boundary problems were considered by Poisson, Ostrogradski, Delaunay, Sarrus, and Cauchy. In 1842 the French Academy proposed as topic for their great mathematical prize the problem to derive the natural boundary conditions which together with Euler's equations must be satisfied by minimizers and maximizers of free boundary value problems for multiple integrals. Four papers were sent in; the prize went to Sarrus with an honourable mentioning of Delaunay, and in 1861 Todhunter [1] held Sarrus's paper for "the most important original contribution to the calculus of variations which has been made during the present century". It is hard to believe that these formulas which can nowadays be derived in a few lines were so highly appreciated by the Academy, but we must realize that in those days integration by parts was not a fully developed tool. This example shows very well how the problems posed by the variational calculus forced analysts to develop new tools. Time and again we find similar examples in the history of this field.

In Chapters 1–4 we have presented all formal aspects of the calculus of variations including all necessary conditions. We have simultaneously treated extrema of single and multiple integrals as there is barely any difference in the degree of difficulty, at least as long as one sticks to variational problems involving only first order derivatives. The difference between one- and multidimensional problems is rarely visible in the formal aspect of the theory but becomes only perceptible when one really wants to construct solutions. This is due to the fact that the necessary conditions for one-dimensional integrals are ordinary differential equations, whereas the Euler equations for multiple integrals are partial differential equations. The problem to solve such equations under prescribed boundary conditions is a much more difficult task than the corresponding problem for ordinary differential equations; except for some special cases it was only solved in this century. As we need rather refined tools of analysis to tackle partial differential equations we deal here only with the formal aspects of the calculus of variations in full generality while existence questions are merely studied for one-dimensional variational problems. The existence and regularity theory of multiple variational integrals will be treated in a separate treatise.

Scheeffer and Weierstrass discovered that positivity of the second variation at a stationary curve is not enough to ensure that the curve furnishes a local minimum; in general one can only show that it is a *weak minimizer*. This means that the curve yields a minimum only in comparison to those curves whose tangents are not much different.

In 1879 Weierstrass discovered a method which enables one to establish a *strong minimum property* for solutions of Euler's equations, i.e. for stationary curves; this method has become known as *Weierstrass field theory*. In essence Weierstrass's method is a rather subtle convexity argument which uses two ingredients. First one employs a local convexity assumption on the integrand of the variational integral which is formulated by means of *Weierstrass's excess function*. Secondly, to make proper use of this assumption one has to embed the given stationary curve in a suitable field of such curves. This field embedding can be interpreted as an introduction of a particular system of normal coordinates which very much simplify the comparison of the given stationary curve with any neighbouring curve. In the plane it suffices to embed the given curve in an arbitrarily chosen field of stationary curves while in higher dimensions one has to embed the curve in a so-called *Mayer field*.

In Chapter 6 of this volume we shall describe Weierstrass field theory for nonparametric one-dimensional variational problems and the contributions of Mayer, Kneser, Hilbert and Carathéodory. The corresponding field theory for parametric integrals is presented in Chapter 8. There we have also a first glimpse at the so-called *direct method* of the calculus of variations. This is a way to establish directly the existence of minimizers by means of set-theoretic arguments; another treatise will entirely be devoted to this subject. In addition we sketch field theories for multiple integrals at the end of Chapters 6 and 7.

In Chapter 7 we describe an important involutory transformation, which will be used to derive a dual picture of the Euler–Lagrange formalism and of field theory, called *canonical formalism*. In this description the dualism *ray versus wave* (or: particle–wave) becomes particularly transparent. The canonical formalism is a part of the *Hamilton–Jacobi theory*, of which we give a self-contained presentation in Chapter 9, together with a brief introduction to symplectic geometry. This theory has its roots in Hamilton's investigations on geometrical optics, in particular on systems of rays. Later Hamilton realized that his formalism is also suited to describe systems of point mechanics, and Jacobi developed this formalism further to an effective integration theory of ordinary and partial differential equations and to a theory of canonical mappings. The connection between *canonical* (or *symplectic*) *transformations* and *Lie's theory of contact transformations* is discussed in Chapter 10 where we also investigate the relations between the principles of Fermat and Huygens. Moreover we treat *Cauchy's method* of integrating partial differential equations of first order by the method of characteristics and illustrate the connection of this technique with Lie's theory.

The reader can use the detailed table of contents with its numerous catchwords as a guideline through the book; the detailed introductions preceding each chapter and also every section and subsection are meant to assist the reader in obtaining a quick orientation. A comprehensive *glimpse at the literature* on the Calculus of Variations is given at the end of Volume 2. Further references can be found in the Scholia to each chapter and in our bibliography. Moreover, important historical references are often contained in footnotes. As important examples are sometimes spread over several sections, we have added a *list of examples*, which the reader can also use to locate specific examples for which he is looking.

Contents of Calculus of Variations II
The Hamiltonian Formalism

Part III. Canonical Formalism and Parametric Variational Problems

Chapter 7. Legendre Transformation, Hamiltonian Systems, Convexity, Field Theories 3

1. Legendre Transformations 4
   1.1. Gradient Mappings and Legendre Transformations 5
        (Definitions. Involutory character of the Legendre transformation. Conjugate convex functions. Young's inequality. Support function. Clairaut's differential equation. Minimal surface equation. Compressible two-dimensional steady flow. Application of Legendre transformations to quadratic forms and convex bodies. Partial Legendre transformations.)
   1.2. Legendre Duality Between Phase and Cophase Space.
        Euler Equations and Hamilton Equations. Hamilton Tensor   18
        (Configuration space, phase space, cophase space, extended configuration (phase, cophase) space. Momenta. Hamiltonians. Energy-momentum tensor. Hamiltonian systems of canonical equations. Dual Noether equations. Free boundary conditions in canonical form. Canonical form of E. Noether's theorem, of Weierstrass's excess function and of transversality.)
2. Hamiltonian Formulation
   of the One-Dimensional Variational Calculus 26
   2.1. Canonical Equations and the Partial Differential Equation
        of Hamilton–Jacobi 26
        (Eulerian flows and Hamiltonian flows as prolongations of extremal bundles. Canonical description of Mayer fields. The 1-forms of Beltrami and Cartan. The Hamilton-Jacobi equation as canonical version of Carathéodory's equations. Lagrange brackets and Mayer bundles in canonical form.)
   2.2. Hamiltonian Flows and Their Eigentime Functions.
        Regular Mayer Flows and Lagrange Manifolds 33
        (The eigentime function of an r-parameter Hamiltonian flow. The Cauchy representation of the pull-back $h^*\kappa_H$ of the Cartan form κ_H with respect to an r-parameter Hamilton flow h by means of an eigentime function. Mayer flows, field-like Mayer bundles, and Lagrange manifolds.)
   2.3. Accessory Hamiltonians and the Canonical Form
        of the Jacobi Equation 41
        (The Legendre transform of the accessory Lagrangian is the accessory Hamiltonian, i.e. the quadratic part of the full Hamiltonian, and its canonical equations describe Jacobi fields. Expressions for the first and second variations.)

2.4. The Cauchy Problem for the Hamilton–Jacobi Equation 48
(Necessary and sufficient conditions for the local solvability of the Cauchy problem. The Hamilton-Jacobi equation. Extension to discontinuous media: refracted light bundles and the theorem of Malus.)

3. Convexity and Legendre Transformations 54
   3.1. Convex Bodies and Convex Functions in $\mathbb{R}^n$ 55
   (Basic properties of convex sets and convex bodies. Supporting hyperplanes. Convex hull. Lipschitz continuity of convex functions.)
   3.2. Support Function, Distance Function, Polar Body 66
   (Gauge functions. Distance function and support function. The support function of a convex body is the distance function of its polar body, and vice versa. The polarity map. Polar body and Legendre transform.)
   3.3. Smooth and Nonsmooth Convex Functions. Fenchel Duality 75
   (Characterization of smooth convex functions. Supporting hyperplanes and differentiability. Regularization of convex functions. Legendre-Fenchel transform.)

4. Field Theories for Multiple Integrals 94
   4.1. DeDonder–Weyl's Field Theory 96
   (Null Lagrangians of divergence type as calibrators. Weyl equations. Geodesic slope fields or Weyl fields, eikonal mappings. Beltrami form. Legendre transformation. Cartan form. DeDonder's partial differential equation. Extremals fitting a geodesic slope field. Solution of the local fitting problem.)
   4.2. Carathéodory's Field Theory 106
   (Carathéodory's involutory transformation, Carathéodory transform. Transversality. Carathéodory calibrator. Geodesic slope fields and their eikonal maps. Carathéodory equations. Vessiot-Carathéodory equation. Generalization of Kneser's transversality theorem. Solution of the local fitting problem for a given extremal.)
   4.3. Lepage's General Field Theory 131
   (The general Beltrami form. Lepage's formalism. Geodesic slope fields. Lepage calibrators.)
   4.4. Pontryagin's Maximum Principle 136
   (Calibrators and pseudonecessary optimality conditions. (I) One-dimensional variational problems with nonholonomic constraints: Lagrange multipliers. Pontryagin's function, Hamilton function, Pontryagin's maximum principle and canonical equations. (II) Pontryagin's maximum principle for multi-dimensional problems of optimal control.)

5. Scholia 146

Chapter 8. Parametric Variational Integrals 153

1. Necessary Conditions 154
   1.1. Formulation of the Parametric Problem. Extremals and Weak Extremals 155
   (Parametric Lagrangians. Parameter-invariant integrals. Riemannian metrics. Finsler metrics. Parametric extremals. Transversality of line elements. Eulerian covector field and Noether's equation. Gauss's equation. Jacobi's variational principle for the motion of a point mass in $\mathbb{R}^3$.)

1.2. Transition from Nonparametric to Parametric Problems and Vice Versa 166
(Nonparametric restrictions of parametric Lagrangians. Parametric extensions of nonparametric Lagrangians. Relations between parametric and nonparametric extremals.)

1.3. Weak Extremals, Discontinuous Solutions, Weierstrass–Erdmann Corner Conditions. Fermat's Principle and the Law of Refraction 171
(Weak D^1- and C^1-extremals. DuBois-Reymond's equation. Weierstrass-Erdmann corner conditions. Regularity theorem for weak D^1-extremals. Snellius's law of refraction and Fermat's principle.)

2. Canonical Formalism and the Parametric Legendre Condition 180

2.1. The Associated Quadratic Problem. Hamilton's Function and the Canonical Formalism 180
(The associated quadratic Lagrangian Q of a parametric Lagrangian F. Elliptic and nonsingular line elements. A natural Hamiltonian and the corresponding canonical formalism. Parametric form of Hamilton's canonical equations.)

2.2. Jacobi's Geometric Principle of Least Action 188
(The conservation of energy and Jacobi's least action principle: a geometric description of orbits.)

2.3. The Parametric Legendre Condition and Carathéodory's Hamiltonians 192
(The parametric Legendre condition or C-regularity. Carathéodory's canonical formalism.)

2.4. Indicatrix, Figuratrix, and Excess Function 201
(Indicatrix, figuratrix and canonical coordinates. Strong and semistrong line elements. Regularity of broken extremals. Geometric interpretation of the excess function.)

3. Field Theory for Parametric Integrals 213

3.1. Mayer Fields and their Eikonals 214
(Parametric fields and their direction fields. Equivalent fields. The parametric Carathéodory equations. Mayer fields and their eikonals. Hilbert's independent integral. Weierstrass's representation formula. Kneser's transversality theorem. The parametric Beltrami form. Normal fields of extremals and Mayer fields, Weierstrass fields, optimal fields, Mayer bundles of extremals.)

3.2. Canonical Description of Mayer Fields 227
(The parametric Cartan form. The parametric Hamilton-Jacobi equation or eikonal equation. One-parameter families of F-equidistant surfaces.)

3.3. Sufficient Conditions 229
(F- and Q-minimizers. Regular Q-minimizers are quasinormal. Conjugate values and conjugate points of F-extremals. F-extremals without conjugate points are local minimizers. Stigmatic bundles of quasinormal extremals and the exponential map of a parametric Lagrangian. F- and Q-Mayer fields. Wave fronts.)

3.4. Huygens's Principle 243
(Complete Figures. Duality between light rays and wave fronts. Huygens's envelope construction of wave fronts. F-distance function. Foliations by one-parameter families of F-equidistant surfaces and optimal fields.)

4. Existence of Minimizers 248
   4.1. A Direct Method Based on Local Existence 248
        (The distance function $d(P, P')$ related to F and its continuity and lower
        semicontinuity properties. Existence of global minimizers based on the local
        existence theory developed in 3.3. Regularity of minimizers.)
   4.2. Another Direct Method Using Lower Semicontinuity 254
        (Minimizing sequences. An equivalent minimum problem. Compactness of
        minimizing sequences. Lower semicontinuity of the variational integral. A
        general existence theorem for obstacle problems. Regularity of minimizers.
        Existence of minimizing F-extremals. Inclusion principle.)
   4.3. Surfaces of Revolution with Least Area 263
        (Comparison of curves with the Goldschmidt polygon. Todhunter's
        ellipse. Comparison of catenaries and Goldschmidt polygons. Conclusive
        results.)
   4.4. Geodesics on Compact Surfaces 270
        (Existence and regularity of F-extremals wich minimize the arc
        length.)
5. Scholia 275

Part IV. Hamilton–Jacobi Theory
and Partial Differential Equations of First Order

Chapter 9. Hamilton–Jacobi Theory and Canonical Transformations . 283

1. Vector Fields and 1-Parameter Flows 288
   1.1. The Local Phase Flow of a Vector Field 290
        (Trajectories, integral curves, maximal flows.)
   1.2. Complete Vector Fields and One-Parameter Groups
        of Transformations 292
        (Infinitesimal transformations.)
   1.3. Lie's Symbol and the Pull-Back of a Vector Field 294
        (The symbol of a vector field and its transformation law.)
   1.4. Lie Brackets and Lie Derivatives of Vector Fields 298
        (Commuting flows. Lie derivative. Jacobi identity.)
   1.5. Equivalent Vector Fields 303
        (Rectification of nonsingular vector fields.)
   1.6. First Integrals 304
        (Time-dependent and time-independent first integrals. Functionally
        independent first integrals. The motion in a central field. Kepler's problem.
        The two-body problem.)
   1.7. Examples of First Integrals 314
        (Lax pairs. Toda lattice.)
   1.8. First-Order Differential Equations
        for Matrix-Valued Functions. Variational Equations.
        Volume Preserving Flows 317
        (Liouville formula. Liouville theorem. Autonomous Hamiltonian flows are
        volume preserving.)
   1.9. Flows on Manifolds 320
        (Geodesics on S^2.)